SAXON MATH™ 3

Math Center Activities

Nancy Larson

A Harcourt Achieve Imprint

www.SaxonPublishers.com
1-800-284-7019

ISBN-13: 978-1-6027-7026-3
ISBN-10: 1-6027-7026-3

Printed in China

9 0940 14 13 12 11 10

Table of Contents

Math Center Overview

What is a math center?

A math center is an area in the classroom where math materials are available for children to use in a small-group setting. A math center may be a table or an area on the floor.

Why have a math center?

A math center provides children with an opportunity to practice skills presented in class, explore materials, work independently, learn how to work with other children, and receive instruction at their instructional level.

A math center provides the teacher with the opportunity to observe children working, assess children's understanding and mastery as they work independently, and meet and work with small groups of children for remediation, reteaching, or enrichment.

How do I set up a math center?

Choose an area of the classroom where you will store the math materials. The children will need a table or room on the floor on which to work.

Materials in the center need to be organized so that children can find them easily and clean up independently. The manipulatives should be stored in baskets or containers that children can access easily. Since the children will be responsible for taking out and putting away the math center materials, the containers should be clearly labeled with the name of the manipulative and a picture, if possible. The shelf where the material is stored should have a label that matches the one on the container.

A material should be put in the math center only after it has been introduced in a lesson. This will ensure that children know how to use the material appropriately.

How do I introduce the math center to the children?

The math center can be introduced after you teach Lesson 1. You will need to show children where the materials are located in the center, where the children may use the materials in the center, and how they will put the materials away when they finish working. Follow the same procedure each time a new math manipulative is introduced in a lesson.

Children will also need to learn how to talk and move in the math center. It is important to demonstrate and then ask children to model how they will walk in the center and how loudly they may speak when they work in the center.

When should I introduce a new math center activity?

A math center activity can be introduced at any time after the lesson is taught: at the end of the lesson, the next day, or a week or two later.

When introducing a new activity, it is important to describe the math center activity to the children, demonstrate the activity in the center, and if necessary, ask a child to model the activity.

Which math center activities should I choose?

Choose activities that will provide children with the independent practice that they most need. For example, if children have difficulty ordering numbers during a lesson, choose a math center activity that will allow children to practice this skill.

Lessons during which only a few children have the chance to use a material (such as a balance) should have a follow-up activity in the math center. This will give all children the chance to use the material individually.

There will also be activities that children are excited about and want to continue at the end of a lesson. They can continue these activities in the math center.

How many math center activities should I use at a time?

You may want to have one to three teacher-directed activities at a time in the math center. When children complete these activities, you may want to make available other previously introduced activities for the children to choose from.

How do I check what children do in a math center?

As children begin a math center task, watch them for a few minutes to make sure they understand the activity and are able to complete it independently.

After children complete a task in the math center, they can use a name card† to label their work. Name cards allow a child to complete a task and go on to another activity while waiting for you to check his or her work when time permits.

It is very important to check each child's work and review incorrectly completed tasks with the child. This can be done while the children are working in the center or later in the day.

How should I group children for math center activities?

The math center activities are designed for groups of four children. For most of the activities, the children should be grouped heterogeneously. The composition of the groups should change during the year so that children have an opportunity to work with many other children in the class.

For remediation, reteaching, and enrichment during center time, children should be grouped by skill level.

How do I use a math center for remediation?

If a group of children has difficulty with an assessment question, select the center activities that provide practice of that concept, and work on those activities in the math center with those children.

If a child has difficulty when a new concept is introduced in a lesson, pair the child with a more capable child during the math center activity. Observe and monitor the children as they work together. At times you may need to pair two children having difficulty and work with those children in the math center.

† Use index cards to make name cards for each child.

How do I use a math center for enrichment?

If a group of very capable children work together, they will often extend a math center activity on their own initiative. Encourage this behavior, and as you monitor the group, ask questions that encourage the children to make observations and extend an activity.

You may also want to allow children to create their own math center activities. For example, when coins are introduced in the lessons, allow the children to set up a "store," price the items, prepare receipts, and act as the cashier as the other children purchase the items using coins.

What materials do I need for the math center activities?

The following chart describes each of the 115 math center activities. This quick reference guide will tell you the objective(s), materials, and directions for each activity.

Many of the materials used in the math centers are the same as those used during the lessons. Suggestions for preparing and storing these materials are provided in the lesson booklets. Other materials needed for the math center activities include blackline Center Masters. Blackline Center Masters are found at the end of this booklet. Make as many copies as you need for your class.

Blackline Center Masters are also available on the Resources and Planner CD.

Why are some activities printed in red?

The Math Center activities printed in red are Learning Palette® activities‡. A Learning Palette Kit with materials for four children is available as a separate purchase at *www.SaxonPublishers.com.*

‡ Learning Palette® is a registered trademark of Learning Wrap-ups Inc.

Math Center Activities and Objectives

***The Math Center Activities printed in red are Learning Palette Activities.**

Activity 1	Identifying a Date on a Calendar (Learning Palette Card 1)
Activity 2	Collecting Data and Creating a Real Graph
Activity 3	Reading a Graph (Learning Palette Card 2)
Activity 4	Comparing Two-Digit Numbers
Activity 5	Ordering and Comparing Two-Digit Numbers
Activity 6	Reading a Venn Diagram Identifying Even and Odd Numbers (Learning Palette Card 3)
Activity 7	Identifying Even and Odd Numbers
Activity 8	Making Congruent Shapes
Activity 9	Adding 10 to a Number (Learning Palette Card 4)
Activity 10	Making a Design With a Given Value Using Pattern Blocks
Activity 11	Rounding a Number to the Nearest Ten
Activity 12	Rounding a Number to the Nearest Ten (Learning Palette Card 5)
Activity 13	Identifying the Missing Addition or Subtraction Fact in a Fact Family (Learning Palette Card 6)
Activity 14	Identifying Polygons
Activity 15	Trading 10s and 1s (Learning Palette Card 7)
Activity 16	Counting Dimes, Nickels, and Pennies
Activity 17	Paying for Items Using Dimes, Nickels, and Pennies (Learning Palette Card 8)
Activity 18	Identifying a Fractional Part of a Whole (Learning Palette Card 9)
Activity 19	Comparing Three-Digit Numbers
Activity 20	Identifying Three-Digit Numbers from Words (Learning Palette Card 10)
Activity 21	Drawing Tally Marks to Represent the Number of Times an Event Occurs
Activity 22	Counting Tally Marks (Learning Palette Card 11)
Activity 23	Estimating the Sum of Two Two-Digit Numbers (Learning Palette Card 12)
Activity 24	Estimating and Measuring Length Using 1-in. Color Tiles and 1-cm Cubes Comparing Units of Length

Activity 25	Ordering and Comparing Three-Digit Numbers
Activity 26	Ordering Three-Digit Numbers (Learning Palette Card 13)
Activity 27	Writing “Some, Some More” and “Some, Some Went Away” Stories
Activity 28	Counting Quarters, Dimes, Nickels, and Pennies
Activity 29	Paying for Items with Bills and Coins (Learning Palette Card 14)
Activity 30	Identifying the Number of Minutes in an Hour or Fractional Part of an Hour (Learning Palette Card 15)
Activity 31	Conducting a Survey Drawing a Pictograph with a Scale of 2
Activity 32	Identifying Three-Digit Numbers from Expanded Form (Learning Palette Card 16)
Activity 33	Identifying the Missing Number in an Addition Problem
Activity 34	Multiplication Facts: Multiplying by 1 and 10
Activity 35	Estimating and Measuring the Capacity of Containers
Activity 36	Comparing Two- and Three-Digit Numbers (Learning Palette Card 17)
Activity 37	Using Comparison Symbols (>, <, and =)
Activity 38	Finding Perimeter (Learning Palette Card 18)
Activity 39	Making a Shape for a Given Perimeter
Activity 40	Reading a Table Adding Two-Digit Numbers (Learning Palette Card 19)
Activity 41	Adding Two-Digit Numbers
Activity 42	Naming Points on a Number Line (Learning Palette Card 20)
Activity 43	Multiplication Facts: Multiplying by 7
Activity 44	Making a Bar Graph With a Scale of 5
Activity 45	Reading Bar Graphs With Scales of 2 or 10 (Learning Palette Card 21)
Activity 46	Writing “Equal-Groups” Stories
Activity 47	Identifying a Number Sentence for a Picture of Equal Groups (Learning Palette Card 22)
Activity 48	Making Symmetrical Designs
Activity 49	Identifying Lines of Symmetry (Learning Palette Card 23)
Activity 50	Identifying a Fractional Part of a Set
Activity 51	Identifying a Fractional Part of a Set (Learning Palette Card 24)
Activity 52	Showing Perfect Squares
Activity 53	Using a Concrete Model to Represent and Order Three-Digit Numbers

Activity 81	Adding Fractions With Like Denominators (Learning Palette Card 38)
Activity 82	Comparing Fractions Using a Picture (Learning Palette Card 39)
Activity 83	Multiplication Facts: Multiplying by 3
Activity 84	Weighing Objects Using Pounds Ordering Objects from Lightest to Heaviest
Activity 85	Subtracting Across Zeros (Learning Palette Card 40)
Activity 86	Telling Time to the Quarter Hour (Learning Palette Card 41)
Activity 87	Identifying Fourths on a Number Line (Learning Palette Card 42)
Activity 88	Multiplication Facts: Multiplying by 4
Activity 89	Creating a Design Using Parallel Line Segments
Activity 90	Identifying the Missing Addend for a Sum of 100 (Learning Palette Card 43)
Activity 91	Making and Counting Change for $1.00
Activity 92	Identifying Change (Learning Palette Card 44)
Activity 93	Identifying the Place Value of a Digit (Learning Palette Card 45)
Activity 94	Identifying the Value of a Digit in a Four-Digit Number (Learning Palette Card 46)
Activity 95	Identifying the Missing Multiplication or Division Fact in a Fact Family (Learning Palette Card 47)
Activity 96	Writing a Division Story
Activity 97	Multiplication Facts: Multiplying by 9
Activity 98	Making a Reflection Across a Line of Symmetry
Activity 99	Identifying a Reflection Across a Line of Symmetry (Learning Palette Card 48)
Activity 100	Finding a Fractional Part of a Set Determining Age
Activity 101	Identifying a Fractional Part of a Set (Learning Palette Card 49)
Activity 102	Making Right, Acute, and Obtuse Triangles
Activity 103	Multiplication Facts: Multiplying by 6
Activity 104	Constructing Geometric Solids From Nets
Activity 105	Matching Solids, Angles, and Parallel Sides to Geometric Terms (Learning Palette Card 50)
Activity 106	Multiplying a Two-Digit Number by a One-Digit Number
Activity 107	Multiplying a Two-Digit Number by a One-Digit Number (Learning Palette Card 51)

Math Center Activities Guide

Math Center Activity	Use After Lesson	Objectives	Materials for Four Children	Teacher Directions
1	Meeting 1	• Identifying a Date on a Calendar	• 4 Learning Palettes • 4 Learning Palette Card 1	Ask each child to match the problems and answers on Learning Palette Card 1.
2	2	• Collecting Data and Creating a Real Graph	• basket of color tiles • Center 2 Master *Make 1 copy for each child in class.*	Pair the children. Ask each child to work with a partner to make a graph using color tiles to show the number of children in the class with the given characteristics. Ask each child to color the bar graph on his/her Center 2 Master and write one question about his/her graph for the children in the class to answer.
3	2	• Reading a Graph	• 4 Learning Palettes • 4 Learning Palette Card 2	Ask each child to match the problems and answers on Learning Palette Card 2.
4	3	• Comparing Two-Digit Numbers	• 2 decks of playing cards with the face cards and the 10s removed. (Leave the Aces in to be used as 1s.)	Pair the children. Ask the partners to sit next to one another, shuffle one of the decks of cards, and divide the cards so that both children have the same number of cards. Ask each child to place his/her cards face down in a stack. Ask each child to turn over the top two cards in his/her stack and arrange those cards to make the largest possible two-digit number. The child with the larger number takes the 4 cards. Play continues until the children have used all the cards in their stacks. The child with the most cards wins the game.

Math Center Activity	Use After Lesson	Objectives	Materials for Four Children	Teacher Directions
5	8	• Ordering and Comparing Two-Digit Numbers	• 2 dot cubes (dice) • Center 5 Master *Make 1 copy for each child in class.*	Pair the children. Ask the partners to take turns rolling a dot cube and recording the number of dots in any one of the boxes on his/her Center 5 Master. When all the boxes are filled, ask each child to order his/her two-digit numbers from least to greatest. The child with the least number and the child with the greatest number score a point. Have the children repeat the activity.
6	9	• Reading a Venn Diagram • Identifying Even and Odd Numbers	• 4 Learning Palettes • 4 Learning Palette Card 3	Ask each child to match the problems and answers on Learning Palette Card 3.
7	9	• Identifying Even and Odd Numbers	• 2 decks of playing cards with the face cards and the 10s removed. (Leave the Aces in to be used as 1s.)	Pair the children. Before beginning the game, ask the partners to decide if they will try to make even or odd two-digit numbers. Ask partners to divide the deck of cards evenly, and ask each child to place his/her cards face down in a stack. Ask each child to turn over the top two cards in his/her stack and try to make the type of number (even or odd) chosen before beginning the game. If the child is successful, he/she keeps the two cards. If not, the two cards are placed in a discard pile. Play continues until the children have used all the cards in their stacks. The child with the most cards wins the game.
8	12	• Making Congruent Shapes	• basket of pattern blocks	Pair the children. Ask each partner to make a shape using no more than ten pattern blocks. Ask his/her partner to make a congruent shape using a different combination of pattern blocks.

Math Center Activity	Use After Lesson	Objectives	Materials for Four Children	Teacher Directions
9	14	• Adding 10 to a Number	• 4 Learning Palettes • 4 Learning Palette Card 4	Ask each child to match the problems and answers on Learning Palette Card 4.
10	15-2	• Making a Design With a Given Value Using Pattern Blocks	• basket of pattern blocks	Pair the children. Ask each child to make a design with a value of $10, $20, or $30, using the values assigned to the pattern blocks in Lesson 15-2. Ask partners to check each other's designs. Optional: Ask children to trace and color their designs.
11	19	• Rounding a Number to the Nearest Ten	• 2 decks of playing cards with the face cards and the 10s removed. (Leave the Aces in to be used as 1s.) • scrap paper	Pair the children. Ask each child to write the multiples of 10 from 10 to 100 on scrap paper. Ask partners to divide the deck of cards evenly and ask each child to place his/her cards face down in a stack. Ask the players to take turns turning over two cards and making a two-digit number. Ask each child to round his/her two-digit number to the nearest 10 and cross this multiple of 10 off his/her paper. Play continues until one child has crossed off all the multiples of 10.
12	19	• Rounding a Number to the Nearest Ten	• 4 Learning Palettes • 4 Learning Palette Card 5	Ask each child to match the problems and answers on Learning Palette Card 5.
13	20-1	• Identifying the Missing Addition or Subtraction Fact in a Fact Family	• 4 Learning Palettes • 4 Learning Palette Card 6	Ask each child to match the problems and answers on Learning Palette Card 6.
14	20-2	• Identifying Polygons	• Polygon Game board (from Lesson 20-2)	Children can continue playing the game introduced in Lesson 20-2.
15	22	• Trading 10s and 1s	• 4 Learning Palettes • 4 Learning Palette Card 7	Ask each child to match the problems and answers on Learning Palette Card 7.

Math Center Activity	Use After Lesson	Objectives	Materials for Four Children	Teacher Directions
16	23	• Counting Dimes, Nickels, and Pennies	• 3 small containers labeled A, B, and C • dimes, nickels, and pennies *Put a collection of dimes, nickels, and pennies in each container. Use fewer than 10 dimes, 10 nickels, and 10 pennies per container. Vary the amounts in the containers.* • 2 cups, each containing 10 dimes, 10 nickels, and 10 pennies • Center 16 Master *Make 1 copy for each child in class.*	Pair the children. Ask each pair of children to sort the coins from one of the containers. Ask the children to record the number of dimes, nickels, and pennies and the total amount of money on their Center 16 Masters. Ask the partners to repeat this activity with the other containers of coins. Have partners work together to show and record 7 different ways of making 20¢ using the dimes, nickels, and/or pennies from one of the cups.
17	23	• Paying for Items Using Dimes, Nickels, and Pennies	• 4 Learning Palettes • 4 Learning Palette Card 8	Ask each child to match the problems and answers on Learning Palette Card 8.
18	24	• Identifying a Fractional Part of a Whole	• 4 Learning Palettes • 4 Learning Palette Card 9	Ask each child to match the problems and answers on Learning Palette Card 9.
19	27	• Comparing Three-Digit Numbers	• 2 decks of playing cards with the face cards and the 10s removed. (Leave the Aces in to be used as 1s.)	Pair the children. Ask partners to divide the deck of cards evenly and each child to place his/her cards face down in a stack. Ask each child to turn over the top three cards in his/her stack and arrange those cards to make the largest possible three-digit number. The child with the largest number takes the 6 cards. Play continues until the children have used all the cards in their stacks. The child with the most cards wins the game.
20	27	• Identifying Three-Digit Numbers from Words	• 4 Learning Palettes • 4 Learning Palette Card 10	Ask each child to match the problems and answers on Learning Palette Card 10.

Math Center Activity	Use After Lesson	Objectives	Materials for Four Children	Teacher Directions
21	30-2	• Drawing Tally Marks to Represent the Number of Times an Event Occurs	• 2 cups with 20 small beans or other small objects in each • 2 Math Folders • scrap paper	Pair the children. Ask one child to gently empty one cup of 20 beans on a hundred number chart on a Math Folder and ask his/her partner to make tally marks to show if the beans landed on odd or even numbers. Ask the children to repeat the activity with the other partner tallying.
22	30-2	• Counting Tally Marks	• 4 Learning Palettes • 4 Learning Palette Card 11	Ask each child to match the problems and answers on Learning Palette Card 11.
23	31	• Estimating the Sum of Two Two-Digit Numbers	• 4 Learning Palettes • 4 Learning Palette Card 12	Ask each child to match the problems and answers on Learning Palette Card 12.
24	32	• Estimating and Measuring Length Using 1-in. Color Tiles and 1-cm Cubes • Comparing Units of Length	• 4 classroom objects of varying lengths • centimeter cubes (10-sticks and unit cubes from set of base ten blocks) • basket of color tiles • Center 24 Master *Make 1 copy for each child in class.*	Pair the children. Ask each pair of children to estimate and measure the length of one of the objects using the color tiles. Ask the children to write the name of the object and the number of color tiles used on their Center 24 Masters. Repeat using the centimeter cubes. Ask the children to repeat the activity with the other objects. Ask the children to compare the units of measure.
25	34	• Ordering and Comparing Three-Digit Numbers	• 2 dot cubes (dice) • Center 25 Master *Make 1 copy for each child in class.*	Pair the children. Ask partners to take turns rolling a dot cube and recording the number of dots in one of his/her boxes on Center 25 Master making three-digit numbers. When all the boxes are filled, ask the children to order their numbers from least to greatest. The child with the least number and the child with the greatest number score a point.
26	34	• Ordering Three-Digit Numbers	• 4 Learning Palettes • 4 Learning Palette Card 13	Ask each child to match the problems and answers on Learning Palette Card 13.

Math Center Activity	Use After Lesson	Objectives	Materials for Four Children	Teacher Directions
27	35-2	• Writing "Some, Some More" and "Some, Some Went Away" Stories	• story paper (2 pieces per child)	Pair the children. Ask each child to write a *some, some more* story and a *some, some went away* story. Ask his/her partner to draw a picture and write a number sentence for each story.
28	36	• Counting Quarters, Dimes, Nickels, and Pennies	• 4 cups of 4 quarters, 10 dimes, 10 nickels, and 10 pennies each • Center 28 Master *Make 1 copy for each child in class.*	Pair the children. Ask each pair of children to take 10 coins from a cup for his/her partner to count. Ask the children to record the number of quarters, dimes, nickels, and pennies and the total amount of money on their Center 28 Masters. Ask the partners to repeat this activity three times. Ask the partners to work together to show and record 7 different ways to make 50¢ using quarters, dimes, and nickels.
29	36	• Paying for Items with Bills and Coins	• 4 Learning Palettes • 4 Learning Palette Card 14	Ask each child to match the problems and answers on Learning Palette Card 14.
30	39	• Identifying the Number of Minutes in an Hour or Fractional Part of an Hour	• 4 Learning Palettes • 4 Learning Palette Card 15	Ask each child to match the problems and answers on Learning Palette Card 15.
31	40-2	• Conducting a Survey • Drawing a Pictograph with a Scale of 2	• Lesson Worksheet 40-2B (from Lesson 40-2)	Allow time for each child to work with his/her partner(s) to complete the survey they started in class and draw a pictograph to represent the results. Ask each group to write three observations about their graph and one question to ask the class on the back of their Lesson Worksheet 40-2B.
32	41	• Identifying Three-Digit Numbers from Expanded Form	• 4 Learning Palettes • 4 Learning Palette Card 16	Ask each child to match the problems and answers on Learning Palette Card 16.

Math Center Activity	Use After Lesson	Objectives	Materials for Four Children	Teacher Directions
33	44	• Identifying the Missing Number in an Addition Problem	• 4 sets of digit cards (from Lesson 34)	Pair the children. Ask each child to use a set of digit cards to show a two-digit addition problem and the sum without showing his/her partner. Ask the child to remove one of the digit cards. Ask the partner to identify the missing digit in his/her partner's problem.
34	45-1	• Multiplication Facts: Multiplying by 1 and 10	• 4 sets of Learning Wrap-Ups®	Allow the children to use Wrap-Up 1 and Wrap-Up 10 to practice multiplying by 1 and 10.
35	45-2	• Estimating and Measuring the Capacity of Containers	• 1 small plastic measuring cup • 5 larger plastic containers of various sizes *Label the containers A, B, C, D, and E.* • water, sand, or rice • funnel • Center 35 Master *Make 1 copy for each child in class.*	Pair the children. Ask each pair to estimate and then to count the number of small cups needed to fill the larger plastic containers. Have children record the answers on their Center 35 Masters. Have the children indicate if the container is closer in size to a quart, liter, or gallon.
36	47	• Comparing Two- and Three-Digit Numbers	• 4 Learning Palettes • 4 Learning Palette Card 17	Ask each child to match the problems and answers on Learning Palette Card 17.
37	47	• Using Comparison Symbols (>, <, and =)	• 2 dot cubes (dice) • Center 37 Master *Make 1 copy for each child in class.*	Pair the children. Ask partners to take turns rolling a dot cube and recording the number of dots in any one of the four spaces in number 1 on Center 37 Master. When all four boxes are filled, each child who has a correct comparison circles *True*. If the comparison is incorrect, the child circles *Not True*. Have the children repeat for 2–4.
38	49	• Finding Perimeter	• 4 Learning Palettes • 4 Learning Palette Card 18	Ask each child to match the problems and answers on Learning Palette Card 18.

Math Center Activity	Use After Lesson	Objectives	Materials for Four Children	Teacher Directions
39	50-2	• Making a Shape for a Given Perimeter	• basket of pattern blocks	Pair the children. Ask each child to make shapes that have perimeters of 8 units and 12 units. Ask each child to check his/her partner's shapes.
40	53	• Reading a Table • Adding Two-Digit Numbers	• 4 Learning Palettes • 4 Learning Palette Card 19	Ask each child to match the problems and answers on Learning Palette Card 19.
41	53	• Adding Two-Digit Numbers	• Lesson Worksheet 52A (from Lesson 52) • Center 41 Master *Make 1 copy for each child in class.* • calculator	Pair the children. Ask each child to select items from Lesson Worksheet 52A and find the total cost on Center 41 Master. Ask his/her partner to check the total using a calculator.
42	54	• Naming Points on a Number Line	• 4 Learning Palettes • 4 Learning Palette Card 20	Ask each child to match the problems and answers on Learning Palette Card 20.
43	55-1	• Multiplication Facts: Multiplying by 7	• 4 sets of Wrap-Ups	Allow the children to use Wrap-Up 7 to practice multiplying by 7.
44	55-2	• Making a Bar Graph With a Scale of 5	• 4 bags, each containing 20–50 objects • Center 44 Master *Make 1 copy for each child in class.*	Pair the children. Ask each child to count the objects from one of the bags and share the total with the other children in the group. Ask the child to work with his/her partner to color the bar graph on Center 44 Master.
45	55-2	• Reading Bar Graphs With Scales of 2 or 10	• 4 Learning Palettes • 4 Learning Palette Card 21	Ask each child to match the problems and answers on Learning Palette Card 21.
46	57	• Writing "Equal-Groups" Stories	• story paper (1 piece per child)	Pair the children. Ask each child to write an equal groups story. Ask his/her partner to draw a picture and write the number sentence for the story.
47	57	• Identifying a Number Sentence for a Picture of Equal Groups	• 4 Learning Palettes • 4 Learning Palette Card 22	Ask each child to match the problems and answers on Learning Palette Card 22.

Math Center Activity	Use After Lesson	Objectives	Materials for Four Children	Teacher Directions
48	58	• Making Symmetrical Designs	• copy paper (3 pieces per child) • scissors	Ask children to fold and cut one of the pieces of paper to make a shape with one line of symmetry. Repeat with two folds for two lines of symmetry and three folds for four lines of symmetry. Ask the children to use a dark crayon to trace the lines of symmetry in the shapes. Display the shapes on the bulletin board.
49	58	• Identifying Lines of Symmetry	• 4 Learning Palettes • 4 Learning Palette Card 23	Ask each child to match the problems and answers on Learning Palette Card 23.
50	61	• Identifying a Fractional Part of a Set	• 4 bags of 4 two-color counters • Center 50 Master *Make 1 copy for each child in class.*	Ask each child to shake 4 counters in his/her hands and gently place them on the desk. Ask the child to record the fractional part that is red on Center 50 Master. Repeat nine more times. Ask the child to color the horizontal bar graph to show how many times each fractional part of a set was red. Ask the children to answer the questions at the bottom of the page.
51	61	• Identifying a Fractional Part of a Set	• 4 Learning Palettes • 4 Learning Palette Card 24	Ask each child to match the problems and answers on Learning Palette Card 24.
52	63	• Showing Perfect Squares	• basket of color tiles • self-stick tags	Ask each child to make two different squares using color tiles and write the number sentence for his/her square on a self-stick tag. Ask the child to place the tags next to the squares.

Math Center Activity	Use After Lesson	Objectives	Materials for Four Children	Teacher Directions
53	64	• Using a Concrete Model to Represent and Order Three-Digit Numbers	• 1 set of digit cards (from Lesson 34) • base ten blocks	Ask each child to select 3 digit cards and make and record the smallest three-digit number possible using his/her digit cards. Ask each child to use the base ten blocks to show his/her number. Ask the children in the group to order the numbers from least to greatest. Ask the children to make the greatest three-digit number and repeat the activity.
54	64	• Identifying a Three-Digit Number Represented by a Picture	• 4 Learning Palettes • 4 Learning Palette Card 25	Ask each child to match the problems and answers on Learning Palette Card 25.
55	64	• Identifying a Four-Digit Number Represented by a Picture	• 4 Learning Palettes • 4 Learning Palette Card 26	Ask each child to match the problems and answers on Learning Palette Card 26.
56	66	• Writing "Some, Some More" Stories With Missing Addends	• story paper (1 piece per child)	Pair the children. Ask each child to write a *some, some more* story with a missing addend. Ask his/her partner to write the number sentence for the story.
57	67	• Subtracting Two-Digit Numbers	• 4 blue index cards with one of the following numbers on each card: 80¢, 72¢, 61¢, and 93¢ • 4 yellow index cards with one of the following numbers on each card: 15¢, 26¢, 37¢, and 49¢ • 2 cups of 10 dimes and 2 cups of 20 pennies • Center 57 Master *Make 1 copy for each child in class.*	Pair the children. Ask one child to choose a blue card. Ask the partner to choose a yellow card. Ask the partners to record both of the amounts on their Center 57 Master and find the difference using the subtraction algorithm. Ask the children to use the coins to check their answer. Ask the child with the blue card to count out the dimes and pennies to show the amount written on his/her card. Ask the child with the yellow card to take away the amount shown on his/her card, trading a dime for 10 pennies as needed.

Math Center Activity	Use After Lesson	Objectives	Materials for Four Children	Teacher Directions
58	67	• Subtracting Two-Digit Numbers Using Mental Computation	• 4 Learning Palettes • 4 Learning Palette Card 27	Ask each child to match the problems and answers on Learning Palette Card 27.
59	70-1	• Multiplication Facts: Multiplying by 2	• 4 sets of Wrap-Ups	Allow the children to use Wrap-Up 2 to practice multiplying by 2.
60	72	• Rounding a Number to the Nearest Hundred	• 2 decks of playing cards with the face cards and the 10s removed. (Leave the Aces in to be used as 1s.) • scrap paper	Pair the children. Ask each child to write the multiples of 100 from 100 to 1000 on scrap paper. Ask partners to divide the deck of cards evenly and each child to place his/her cards face down in a stack. Ask the players to take turns turning over three cards and making a three-digit number. Ask each child to round his/her three-digit number to the nearest 100 and cross that multiple of 100 off his/her paper. Play continues until one child has crossed off all the multiples of 100.
61	72	• Rounding a Number to the Nearest Hundred	• 4 Learning Palettes • 4 Learning Palette Card 28	Ask each child to match the problems and answers on Learning Palette Card 28.
62	72	• Estimating Sums of Two Three-Digit Numbers	• 4 Learning Palettes • 4 Learning Palette Card 29	Ask each child to match the problems and answers on Learning Palette Card 29.
63	72	• Estimating the Differences of Three-Digit Numbers	• 4 Learning Palettes • 4 Learning Palette Card 30	Ask each child to match the problems and answers on Learning Palette Card 30.
64	74	• Representing a Fraction Using a Picture	• 4 Learning Palettes • 4 Learning Palette Card 31	Ask each child to match the problems and answers on Learning Palette Card 31.

Math Center Activity	Use After Lesson	Objectives	Materials for Four Children	Teacher Directions
65	76	• Adding Three-Digit Numbers	• 2 dot cubes (dice) • Center 65 Master *Make 1 copy for each child in class.*	Pair the children. Ask partners to take turns rolling a dot cube and recording the number of dots in one of the boxes on his/her Center 65 Master. When all the boxes are filled, ask each child to add his/her three-digit numbers. The child with the greater sum scores a point.
66	78	• Writing Money Amounts Using Words	• Center 66 Master *Make 2–3 copies for each child in class. Cut apart the checks and staple together to make individual "check books."* • pictures of 6 toys with prices and a store name attached	Pair the children. Ask each child to write two checks to purchase two of the toys. Ask his/her partner to check the accuracy of the check. Save the unused checks for Math Center Activity 69.
67	80-2	• Reading Bar Graphs with Scales of 2 or 5	• 4 Learning Palettes • 4 Learning Palette Card 32	Ask each child to match the problems and answers on Learning Palette Card 32.
68	80-2	• Making Reasonable Predictions by Collecting and Analyzing Data	• bags of color tiles (from Lesson 80-2) • Center 68 Master *Make 1 copy for each child in class.*	Pair the children. Ask the children to repeat the activity introduced during Lesson 80-2 using a different bag of color tiles.
69	82	• Adding Money Amounts (Decimals) • Writing Money Amounts Using Words	• Center 66 Master • stapled "check books" (from Math Center Activity 66) • 4–6 catalogs or pictures of toys (from Math Center Activity 66)	Pair the children. Ask each child to choose two or three items from a catalog, find the total cost of the items, and write a check for the cost. Ask his/her partner to check the accuracy of the check.
70	83	• Reading a Thermometer	• 4 Learning Palettes • 4 Learning Palette Card 33	Ask each child to match the problems and answers on Learning Palette Card 33.
71	84	• Identifying Equivalent Units of Time	• 4 Learning Palettes • 4 Learning Palette Card 34	Ask each child to match the problems and answers on Learning Palette Card 34.
72	85-1	• Multiplication Facts: Multiplying by 5	• 4 sets of Wrap-Ups	Allow the children to use Wrap-Up 5 to practice multiplying by 5.

Math Center Activity	Use After Lesson	Objectives	Materials for Four Children	Teacher Directions
73	85-2	• Measuring Distance Using Feet, Yards, and Meters	• adding machine tape (10-foot piece per pair of children) • rulers • meterstick • 2 red, 2 blue, and 2 black markers • 6 objects • Center 73 Master *Make 1 copy for each child in class.*	Pair the children. Ask each pair to make a measuring tape by using a ruler and a blue marker to mark off feet and a black marker to mark off yards on one side of a piece of adding machine tape; and a meterstick and a red marker to mark off meters on the other side of the tape. Ask the pairs of children to use the measuring tape to measure the distance between three pairs of objects and record the distances on their Center 73 Masters.
74	87	• Making an Array for a Number Sentence • Identifying a Number Sentence for an Array	• 8 index cards *Write one of the following on each of the index cards: 2 x 3; 3 x 2; 2 x 4; 4 x 2; 3 x 4; 4 x 3; 2 x 5; 5 x 2* *Draw and label each array on the back of each card.* • basket of color tiles • scrap paper	Ask each child to choose a card and use color tiles to make the array. When the child finishes, ask the child to turn over the card to the array. Repeat with three more cards.
75	87	• Identifying a Number Sentence for an Array	• 4 Learning Palettes • 4 Learning Palette Card 35	Ask each child to match the problems and answers on Learning Palette Card 35.
76	88	• Finding Area by Counting Square Units	• 6 construction paper rectangles with the following dimensions: 3" x 6"; 2" x 6"; 4" x 4"; 2" x 5"; 2" x 4"; 3" x 5" *Label the construction paper rectangles with the letters A, B, C, D, E, and F.* • baskets of color tiles • Center 76 Master *Make 1 copy for each child in class.*	Ask each child to predict the order of the rectangles from smallest to largest in area. Ask each child to use the color tiles to find the area of the rectangles. Ask the child to record the area of each rectangle on his/her Center 76 Master and to order the rectangles from smallest to largest in area.
77	88	• Making Rectangles With a Given Area	• basket of color tiles • Center 77 Master *Make 1 copy for each child in class.*	Ask each child to make all possible rectangles with an area of 24 square units. Ask the child to record the rectangles on Center 77 Master.

Math Center Activity	Use After Lesson	Objectives	Materials for Four Children	Teacher Directions
78	88	• Finding the Area of a Rectangle	• 4 Learning Palettes • 4 Learning Palette Card 36	Ask each child to match the problems and answers on Learning Palette Card 36.
79	90-2	• Creating a Spinner • Determining the Likelihood of an Event	• 4 paper clips (from Lesson 90-2) • Center 79 Master *Make 1 copy for each child in class.*	Ask each child to make a spinner on his/her Center 79 Master using 4 colors. Ask each child to predict the color the spinner is more likely, less likely, and equally likely to land on. Ask the child to record the results of spinning the spinner.
80	90-2	• Identifying the Likelihood of an Event	• 4 Learning Palettes • 4 Learning Palette Card 37	Ask each child to match the problems and answers on Learning Palette Card 37.
81	93	• Adding Fractions With Like Denominators	• 4 Learning Palettes • 4 Learning Palette Card 38	Ask each child to match the problems and answers on Learning Palette Card 38.
82	94	• Comparing Fractions Using a Picture	• 4 Learning Palettes • 4 Learning Palette Card 39	Ask each child to match the problems and answers on Learning Palette Card 39.
83	95-1	• Multiplication Facts: Multiplying by 3	• 4 sets of Wrap-Ups	Allow the children to use Wrap-Up 3 to practice multiplying by 3.
84	95-2	• Weighing Objects Using Pounds • Ordering Objects from Lightest to Heaviest	• bathroom scale • 5 objects between 2 and 10 pounds that will fit easily on the bathroom scale • Center 84 Master *Make 1 copy for each child in class.*	Pair children. Ask each pair to predict the order of the objects by weight from lightest to heaviest. Ask each pair to weigh the objects on the scale and write the names of the objects and their weights on their Center 84 Masters. Ask each pair to put the objects in order from lightest to heaviest.
85	96	• Subtracting Across Zeros	• 4 Learning Palettes • 4 Learning Palette Card 40	Ask each child to match the problems and answers on Learning Palette Card 40.
86	97	• Telling Time to the Quarter Hour	• 4 Learning Palettes • 4 Learning Palette Card 41	Ask each child to match the problems and answers on Learning Palette Card 41.
87	99	• Identifying Fourths on a Number Line	• 4 Learning Palettes • 4 Learning Palette Card 42	Ask each child to match the problems and answers on Learning Palette Card 42.

Math Center Activity	Use After Lesson	Objectives	Materials for Four Children	Teacher Directions
88	100-1	• Multiplication Facts: Multiplying by 4	• 4 sets of Wrap-Ups	Allow the children to use Wrap-Up 4 to practice multiplying by 4.
89	100-2	• Creating a Design Using Parallel Line Segments	• construction paper • rulers • markers	Ask each child to create a picture using only line segments. The picture should have at least five sets of parallel line segments. Ask other children to identify the parallel line segments in the picture.
90	101	• Identifying the Missing Addend for a Sum of 100	• 4 Learning Palettes • 4 Learning Palette Card 43	Ask each child to match the problems and answers on Learning Palette Card 43.
91	102	• Making and Counting Change for $1.00	• 8 items with price tags (from Lesson 102) • 2 cups with 3 quarters, 5 dimes, 5 nickels, and 10 pennies in each • Lesson Worksheet 102B (from Lesson 102)	Pair children. Ask each child to select an object and show the change he/she will receive if he/she pays for the object with a one-dollar bill. Ask the children to take turns counting back the change to their partner. Ask the children to repeat the activity with different objects.
92	102	• Identifying Change	• 4 Learning Palettes • 4 Learning Palette Card 44	Ask each child to match the problems and answers on Learning Palette Card 44.
93	103	• Identifying the Place Value of a Digit	• 4 Learning Palettes • 4 Learning Palette Card 45	Ask each child to match the problems and answers on Learning Palette Card 45.
94	104	• Identifying the Value of a Digit in a Four-Digit Number	• 4 Learning Palettes • 4 Learning Palette Card 46	Ask each child to match the problems and answers on Learning Palette Card 46.
95	105-1	• Identifying the Missing Multiplication or Division Fact in a Fact Family	• 4 Learning Palettes • 4 Learning Palette Card 47	Ask each child to match the problems and answers on Learning Palette Card 47.
96	108	• Writing a Division Story	• story paper (1 piece per child)	Pair the children. Ask each child to write a division story. Ask his/her partner to draw a picture and write the number sentence for the story.

Math Center Activity	Use After Lesson	Objectives	Materials for Four Children	Teacher Directions
97	110-1	• Multiplication Facts: Multiplying by 9	• 4 sets of Wrap-Ups	Allow the children to use Wrap-Up 9 to practice multiplying by 9.
98	110-2	• Making a Reflection Across a Line of Symmetry	• 8 bags, each containing 3 of each color pattern block • 4 pieces of white construction paper with a black line drawn down the center	Ask each child to use one bag of pattern blocks to make a design on one side of the line so that the design is touching the black line. After each child makes a design, ask him/her to use another bag of pattern blocks to make a reflection on the other side of the black line so that the design will be symmetrical.
99	110-2	• Identifying a Reflection Across a Line of Symmetry	• 4 Learning Palettes • 4 Learning Palette Card 48	Ask each child to match the problems and answers on Learning Palette Card 48.
100	111	• Finding a Fractional Part of a Set • Determining Age	• 4 cups of 10 pennies each • Center 100 Master *Make 1 copy for each child in class.*	Ask each child to order the pennies in his/her cup by year. After ordering the pennies, ask each child to divide the pennies into three groups—pennies that are older than the child, pennies that are the same age as the child, and pennies that are younger than the child. Have the child count the pennies in each of the three groups and record the fractional part of the pennies that are older, the same age, and younger than the child on his/her Center 100 Master.
101	111	• Identifying a Fractional Part of a Set	• 4 Learning Palettes • 4 Learning Palette Card 49	Ask each child to match the problems and answers on Learning Palette Card 49.

Math Center Activity	Use After Lesson	Objectives	Materials for Four Children	Teacher Directions
102	113	• Making Right, Acute, and Obtuse Triangles	• construction paper • rulers • scissors	Ask each child to draw three triangles—one acute, one right, and one obtuse—on a piece of construction paper. Have each child cut out his/her triangles. Ask each child to label the angles in his/her triangles and to write his/her name on each triangle. Display the children's triangles on a bulletin board.
103	115-1	• Multiplication Facts: Multiplying by 6	• 4 sets of Wrap-Ups	Allow the children to use Wrap-Up 6 to practice multiplying by 6.
104	115-2	• Constructing Geometric Solids From Nets	• Teacher Master 115 (from Lesson 115-2) *Make 1 copy for each child in class. (Enlarge these nets to 200%, if possible.)* • scissors • tape	Ask each child to cut out and tape each net to make a pyramid, a triangular prism, and a rectangular prism.
105	115-2	• Matching Solids, Angles, and Parallel Sides to Geometric Terms	• 4 Learning Palettes • 4 Learning Palette Card 50	Ask each child to match the problems and answers on Learning Palette Card 50.
106	116	• Multiplying a Two-Digit Number by a One-Digit Number	• 2 dot cubes (dice) • Center 106 Master *Make 1 copy for each child in class.*	Pair the children. Ask partners to take turns rolling a dot cube and recording the number of dots in one of the boxes on their Center 106 Master. When three boxes are filled, ask each child to multiply his/her one-digit and two-digit numbers. When the children have finished multiplying, ask them to add the three products. The child with the greater sum scores a point.
107	116	• Multiplying a Two-Digit Number by a One-Digit Number	• 4 Learning Palettes • 4 Learning Palette Card 51	Ask each child to match the problems and answers on Learning Palette Card 51.
108	117	• Identifying the Missing Numbers in a Function	• 4 Learning Palettes • 4 Learning Palette Card 52	Ask each child to match the problems and answers on Learning Palette Card 52.

Math Center Activity	Use After Lesson	Objectives	Materials for Four Children	Teacher Directions
109	120-1	• Multiplication Facts: Multiplying by 8	• 4 sets of Wrap-Ups	Allow the children to use Wrap-Up 8 to practice multiplying by 8.
110	120-1	• Identifying the Missing Numbers on a Multiplication Chart	• 4 Learning Palettes • 4 Learning Palette Card 53	Ask each child to match the problems and answers on Learning Palette Card 53.
111	121	• Finding the Volume of a Rectangular Prism	• 3 small boxes labeled A, B, and C • unit cubes (from base-ten blocks) • Center 111 Master *Make 1 copy for each child in class.*	Pair the children. Ask the children to estimate the volume of each small box. Ask the children to use the unit cubes to find the volume of each box and record the results on Center 111 Master.
112	121	• Finding the Volume by Counting Cubic Units	• 4 Learning Palettes • 4 Learning Palette Card 54	Ask each child to match the problems and answers on Learning Palette Card 54.
113	124	• Dividing a Two-Digit Number by a One-Digit Number	• 4 Learning Palettes • 4 Learning Palette Card 55	Ask each child to match the problems and answers on Learning Palette Card 55.
114	128	• Adding Positive and Negative Numbers	• 2 decks of playing cards with the face cards removed. (Leave the Aces in to be used as 1s.) • copies of Lesson Worksheet 128 (from Lesson 128)	Pair the children. Ask partners to divide the deck of cards evenly and ask each child to place his/her cards face down in a stack. Red cards represent negative numbers and black cards represent positive numbers. Ask each child to turn over two cards and find the sum. The child with the greater sum takes the 4 cards. If the sums are equal, each child keeps his/her own cards. Play continues until the children have used all the cards in their stacks. The child with the most cards collected wins the game.
115	129	• Locating Points on a Coordinate Graph	• 4 Learning Palettes • 4 Learning Palette Card 56	Ask each child to match the problems and answers on Learning Palette Card 56.

Name ______________________________

Center 2 Master

***Saxon** Math 3 (for use after **Lesson 2**)*

Color the graph to show one of the following:

How many children are wearing short sleeve or long sleeve shirts?
How many children are wearing shirts with red or shirts without red?
How many children are wearing shirts with buttons or no buttons?

Title ____________________

Question: ______________________________

____

Name ______________________________

Center 2 Master

***Saxon** Math 3 (for use after **Lesson 2**)*

Color the graph to show one of the following:

How many children are wearing short sleeve or long sleeve shirts?
How many children are wearing shirts with red or shirts without red?
How many children are wearing shirts with buttons or no buttons?

Title ____________________

Question: ______________________________

____

Name ________________________________

Center 5 Master

***Saxon** Math 3 (for use after **Lesson 8**)*

______ ______ ______ ______ ______

Least Greatest

This page may be photocopied for educational use within each purchasing institution.

Name ________________________________

Center 5 Master

***Saxon** Math 3 (for use after **Lesson 8**)*

______ ______ ______ ______ ______

Least Greatest

This page may be photocopied for educational use within each purchasing institution.

© Harcourt Achieve Inc. and Nancy Larson. All rights reserved.

Name ____________________________________

Center 16 Master

***Saxon** Math 3 (for use after **Lesson 23**)*

Count the dimes, nickels, and pennies

Count the money.

	dimes	nickels	pennies	
A				________ ¢
B				________ ¢
C				________ ¢

Show 7 different ways to make **20¢** using dimes, nickels, and/or pennies.

________________________________ ________________________________

________________________________ ________________________________

________________________________ ________________________________

____

Name ______________________________

***Saxon** Math 3 (for use after **Lesson 32**)*

Measure four objects using 1-inch color tiles and centimeter cubes.

Objects	1-Inch Color Tiles	Centimeter Cubes

How do the units of measure compare?

____

____

____

____

Name ______________________

Center 25 Master

Saxon Math 3 (for use after Lesson 34)

______ ______ ______ ______ ______

Least Greatest

Name ______________________

Center 25 Master

Saxon Math 3 (for use after Lesson 34)

______ ______ ______ ______ ______

Least Greatest

Name ______________________________

Center 28 Master

***Saxon** Math 3 (for use after **Lesson 36**)*

Count the quarters, dimes, nickels, and pennies

Count the money.

	quarters	dimes	nickels	pennies	
A					______ ¢
B					______ ¢
C					______ ¢

Show 7 different ways to make **50¢** using quarters, dimes, and nickels.

______________________________ ______________________________

______________________________ ______________________________

______________________________ ______________________________

____

Name ______________________________

***Saxon** Math 3 (for use after **Lesson 45-2**)*

Estimate and measure the capacity of the containers.

Container	Estimated Number of Cups	Actual Number of Cups

Write the names of the containers in order from least to greatest capacity.

__________ __________ __________ __________ __________

Least   Greatest

Which container is closest to one quart? ______________________________

Which container is closest to one liter? ______________________________

Which container is closest to one gallon? ______________________________

Name ______________________________

Center 37 Master

***Saxon** Math 3 (for use after **Lesson 47**)*

1. ☐☐ < ☐☐ True Not True

2. ☐☐ < ☐☐ True Not True

3. ☐☐ > 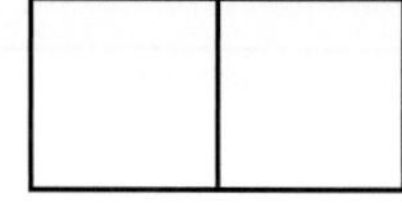 True Not True

4. ☐☐ > 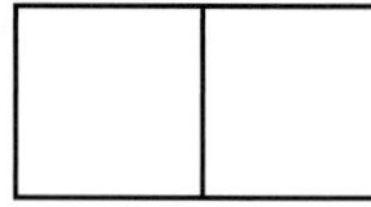 True Not True

Name ______________________________

Center 37 Master

***Saxon** Math 3 (for use after **Lesson 47**)*

1. ☐☐ < 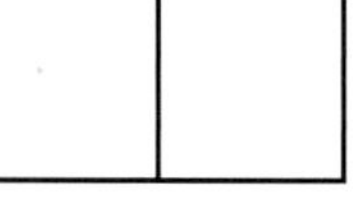 True Not True

2. ☐☐ <  True Not True

3. ☐☐ > 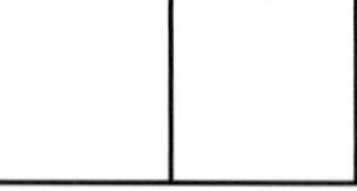 True Not True

4. ☐☐ >  True Not True

This page may be photocopied for educational use within each purchasing institution.

Name ________________________________

***Saxon** Math 3 (for use after **Lesson 53**)*

Lunch Menu Choices

Monday

Meal ____________ ____________

Drink ____________ + ____________

Total ____________

Tuesday

Meal ____________ ____________

Drink ____________ + ____________

Total ____________

Wednesday

Meal ____________ ____________

Drink ____________ ____________

Dessert ____________ + ____________

Total ____________

Thursday

Meal ____________ ____________

Drink ____________ ____________

Dessert ____________ + ____________

Total ____________

Friday

Meal ____________ ____________

Drink ____________ ____________

Dessert ____________ + ____________

Total ____________

Name ____________________

Capacity of a Small Plastic Bag

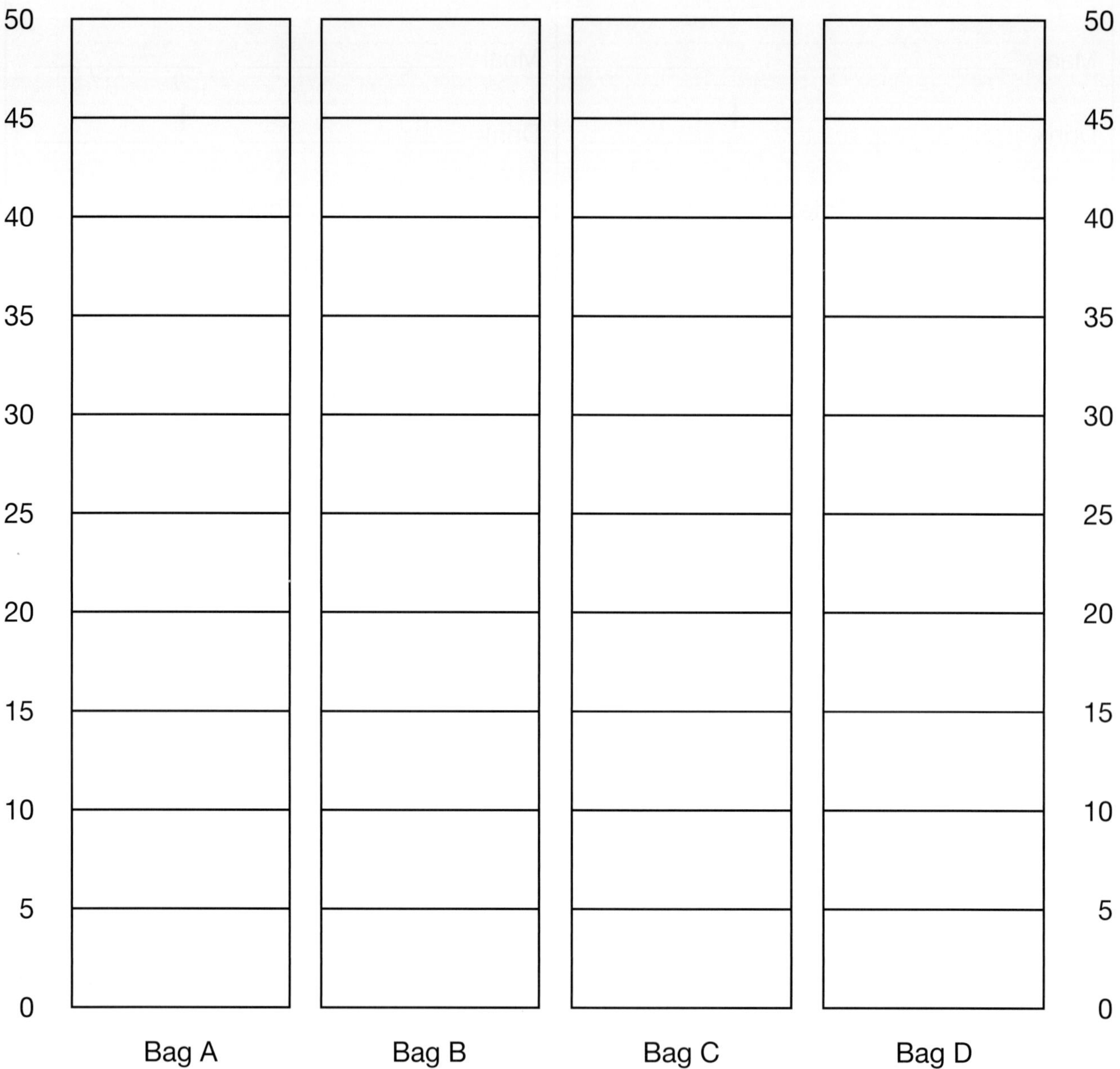

Name ______________________________

A. Shake 4 two-color counters 10 times.
Write what fractional part was red.

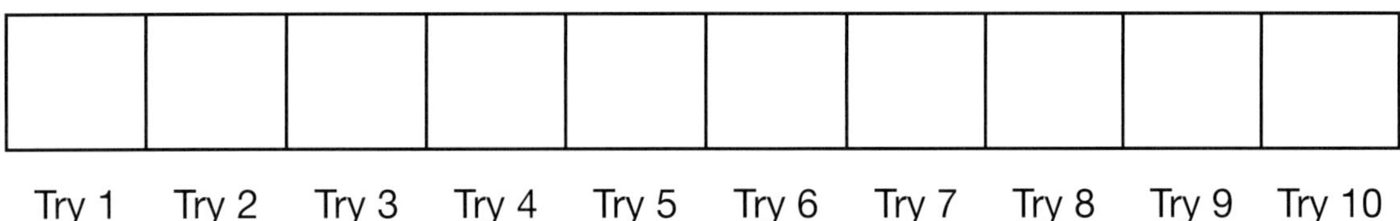

B. Make a horizontal bar graph to show how many times each fractional part of the set was red.

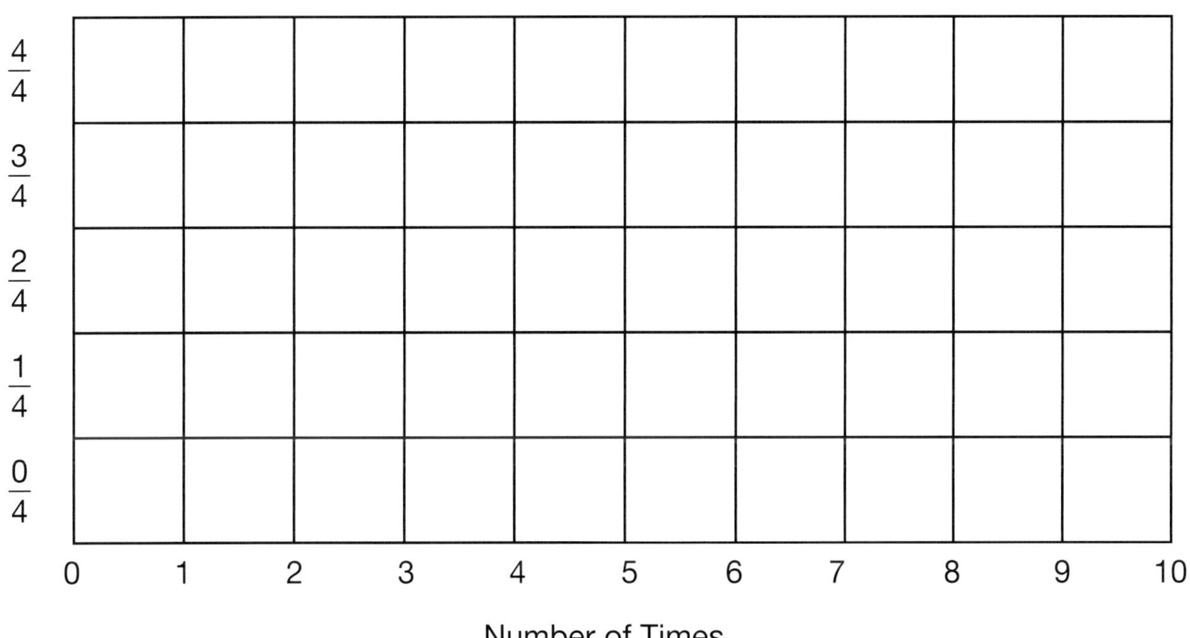

If you did this again, according to this graph, what fractional part of a set are you more likely to get? ______________

If you did this again, according to this graph, what fractional part of a set are you less likely to get? ______________

Name ______________________________

Center 57 Master

***Saxon** Math 3 (for use after **Lesson 67**)*

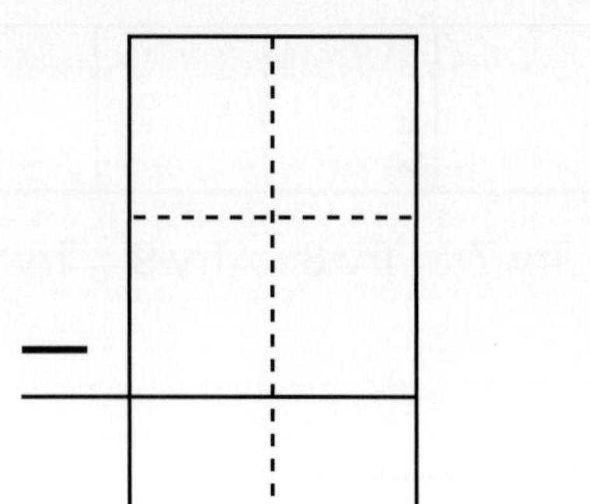

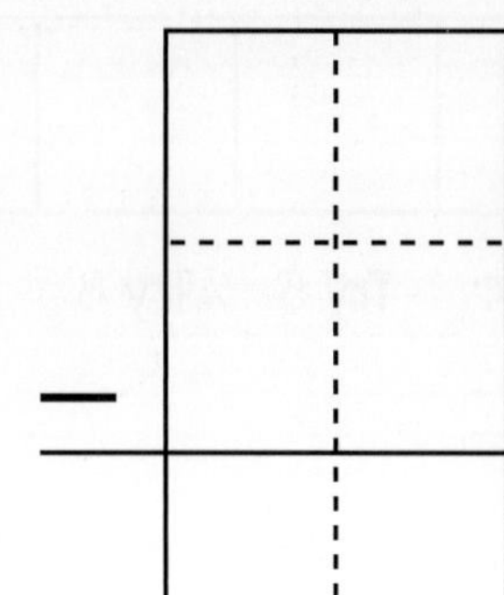

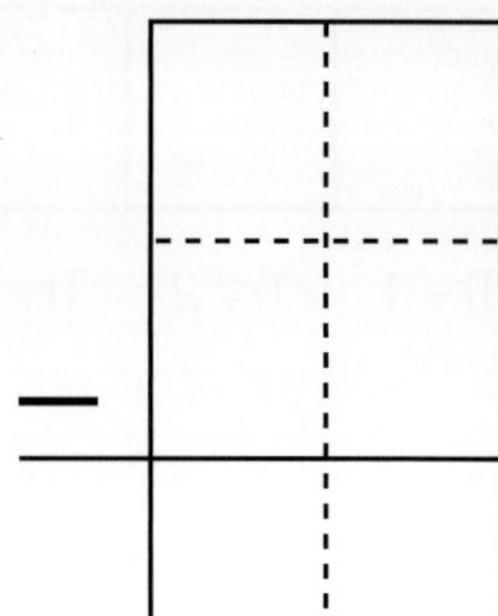

Name ______________________________

Center 57 Master

***Saxon** Math 3 (for use after **Lesson 67**)*

This page may be photocopied for educational use within each purchasing institution.

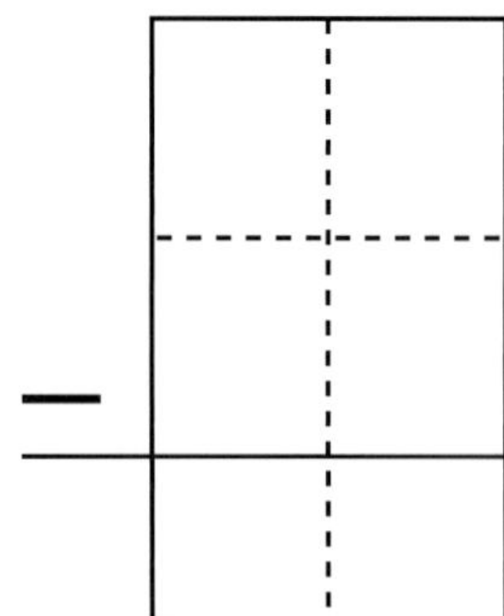

© Harcourt Achieve Inc. and Nancy Larson. All rights reserved.

Name ______________________________

***Saxon** Math 3 (for use after **Lesson 76**)*

1.

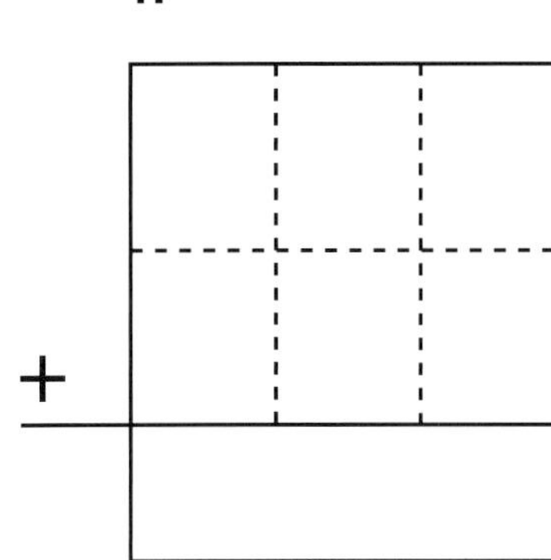

2.

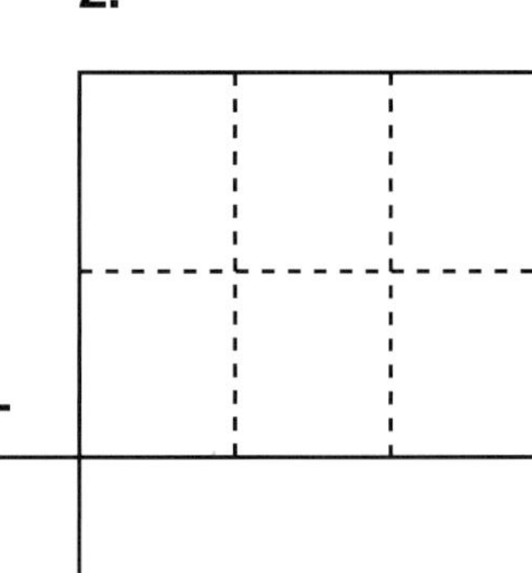

3.

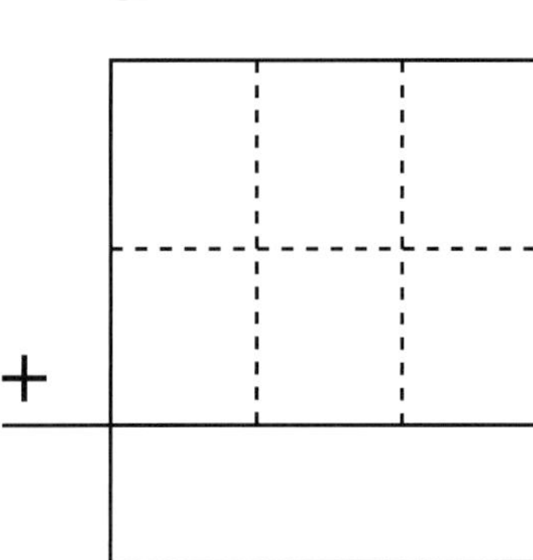

Name ______________________________

***Saxon** Math 3 (for use after **Lesson 76**)*

1.

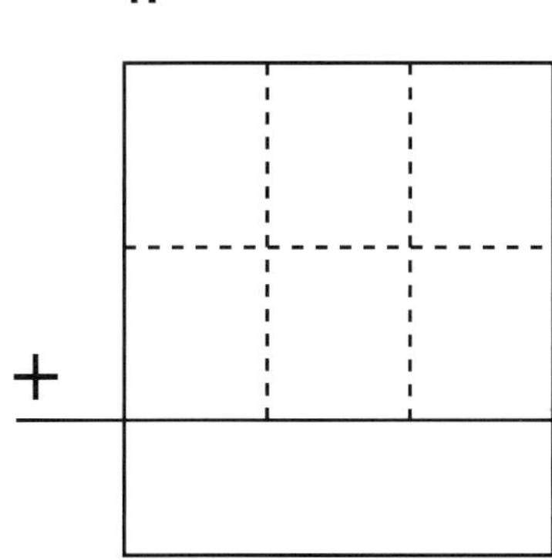

2.

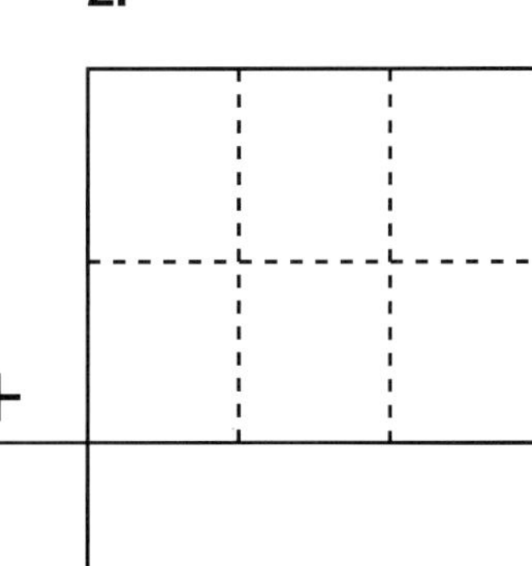

3.

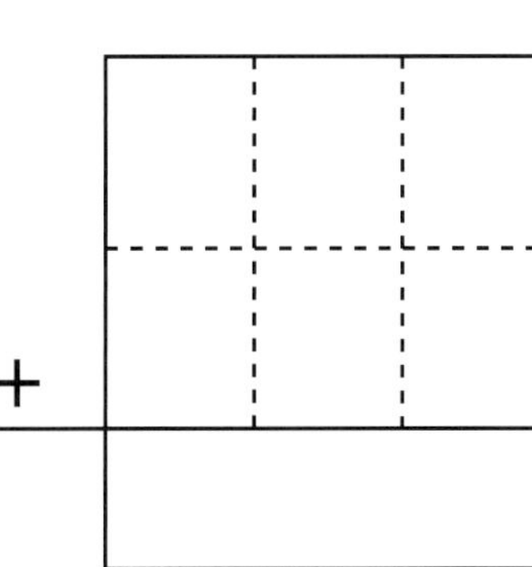

Name ______________________________

Center 66 Master

***Saxon** Math 3 (for use after **Lesson 78**)*

Date ______________________

PAY TO THE
ORDER OF ______________________________ $ [ ]

______________________________ Dollars

____

Date ______________________

PAY TO THE
ORDER OF ______________________________ $ [ ]

______________________________ Dollars

____

Date ______________________

PAY TO THE
ORDER OF ______________________________ $ [ ]

______________________________ Dollars

____

Name ______________________________

Center **68** Master

Saxon *Math 3 (for use after* ***Lesson 80-2****)*

Bag ____

1. Do not look inside the bag.
Take one tile out of the bag.
Tally the color of the tile you drew.
Put the tile back in the bag and mix the tiles.
Do this 80 times.

Red	
Yellow	
Blue	
Green	

2. Make a bar graph showing the number of color tiles you tallied.

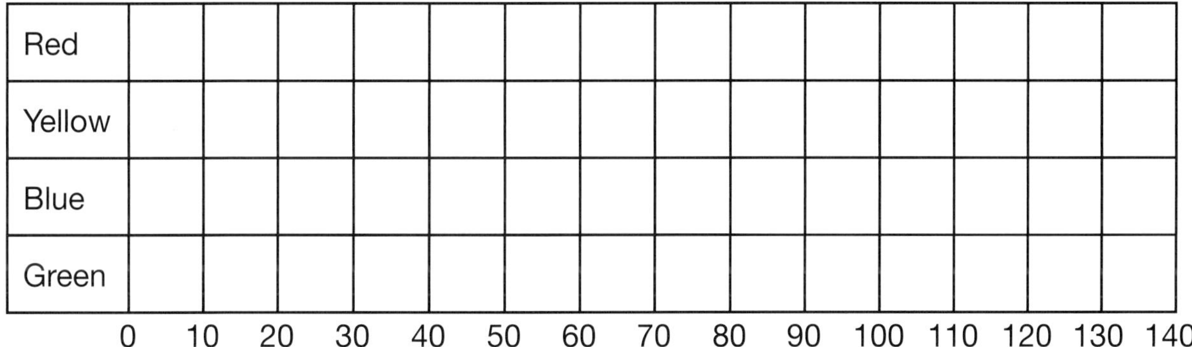

Red														
Yellow														
Blue														
Green														

3. There are 6 tiles in the bag. How many tiles of each color do you think are in the bag?

______ Red ______ Yellow ______ Blue ______ Green

Explain why you think your predictions are reasonable. ______________________________

____

____

____

Name ______________________________

Center 73 Master

***Saxon** Math 3 (for use after **Lesson 85-2**)*

Measure the distance between three objects using yards and feet and meters.

Distances Measured	Yards and Feet	Meters
__________ to __________		
__________ to __________		
__________ to __________		

Name ______________________________

Center 76 Master

***Saxon** Math 3 (for use after **Lesson 88**)*

Predict the order of the rectangles from smallest to largest in area.

________ ________ ________ ________ ________ ________

Smallest Largest

Use color tiles to find the area in square inches of each rectangle.

Rectangle	Area in Square Inches

List the rectangles in order from smallest to largest.

________ ________ ________ ________ ________ ________

Smallest Largest

Name ________________________________

Center 77 Master

***Saxon** Math 3 (for use after **Lesson 88**)*

Name ___________________________________

Saxon Math 3 (for use after **Lesson 90-2**)

Use red, yellow, blue, and green to make a spinner. Color the graph to record the results of 20 spins.

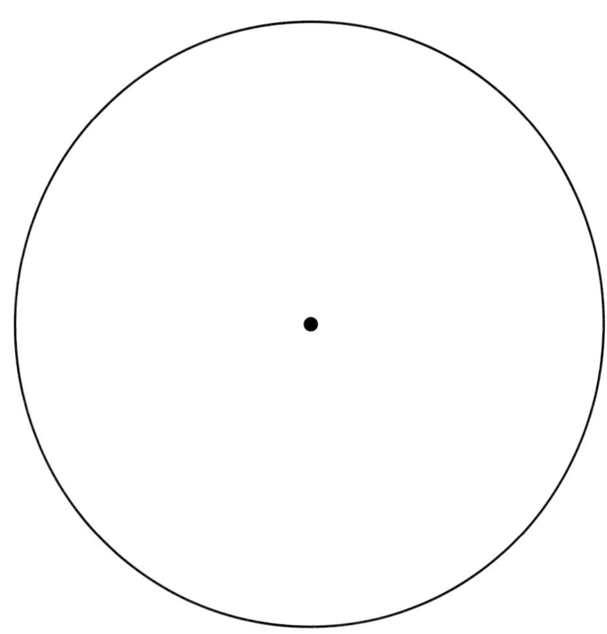

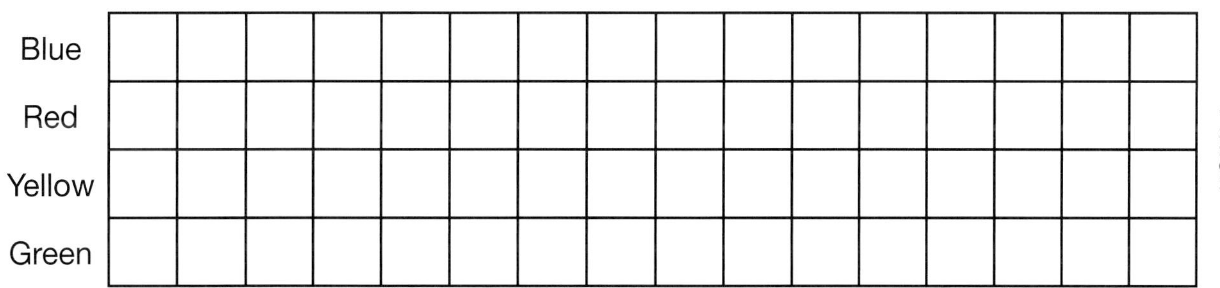

Describe your results. Predict what will happen if you spin 20 more times.

____

____

____

____

Name ______________________________

Center 84 Master

***Saxon** Math 3 (for use after **Lesson 95-2**)*

Predict the order of the objects by weight.

________ ________ ________ ________ ________

Lightest Heaviest

Weigh the objects using pounds.

Objects	Pounds

Write the names of the objects in order from lightest to heaviest.

________ ________ ________ ________ ________

Lightest Heaviest

Name ________________________________

Center 100 Master

Saxon *Math 3 (for use after* ***Lesson 111****)*

List the dates on 10 pennies from oldest to newest.

_______ _______ _______ _______ _______ _______ _______ _______ _______ _______

What fractional part of the pennies are older than you are? _______

What fractional part of the pennies are the same age as you are? _______

What fractional part of the pennies are younger than you are? _______

Name ________________________________

Center 100 Master

Saxon *Math 3 (for use after* ***Lesson 111****)*

List the dates on 10 pennies from oldest to newest.

_______ _______ _______ _______ _______ _______ _______ _______ _______ _______

What fractional part of the pennies are older than you are? _______

What fractional part of the pennies are the same age as you are? _______

What fractional part of the pennies are younger than you are? _______

Name ______________________

Center 106 Master

***Saxon** Math 3 (for use after **Lesson 116**)*

1. ☐ × ☐☐ = ______

2. ☐ × ☐☐ = ______

3. ☐ × ☐☐ = ______

Name ______________________

Center 106 Master

***Saxon** Math 3 (for use after **Lesson 116**)*

1. ☐ × ☐☐ = ______

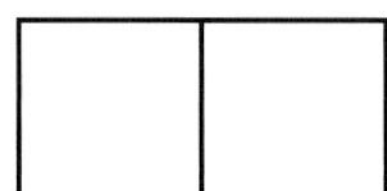

2. ☐ × ☐☐ = ______

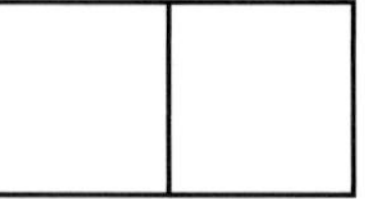

3. ☐ × ☐☐ = ______

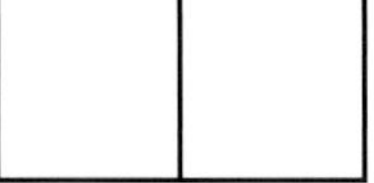

This page may be photocopied for educational use within each purchasing institution.

© Harcourt Achieve Inc. and Nancy Larson. All rights reserved.

Name ______________________________

*Saxon Math 3 (for use after **Lesson 121**)*

Find the volume of each rectangular prism.

	Estimate	Actual
A		

	Estimate	Actual
B		

	Estimate	Actual
C		